FIREFIGHTER

By Diane Lindsey Reeves

Ferguson Publishing
An imprint of Infobase Publishing

Virtual Apprentice: Firefighter

Ferguson
An imprint of Infobase Publishing, Inc.
132 West 31st Street
New York, NY 10001

ISBN-10: 0-8160-6751-1

ISBN-13: 978-0-8160-6751-0

Library of Congress Cataloging-in-Publication Data

Reeves, Diane Lindsey, 1959-
  Virtual apprentice: firefighter / Diane Lindsey Reeves.
      p. cm.
  Includes bibliographical references and index.
  ISBN 0-8160-6751-1 (hc : alk. paper)
  1. Fire extinction–Vocational guidance. 2. Fire extinction–Computer network resources. I. Title.
  TH9119.R443 2006
  628.9'25023–dc22

                                    2006007968

Ferguson books are available at special discounts when purchased in bulk quantities for businesses, associations, institutions, or sales promotions. Please call our Special Sales Department in New York at (212) 967-8800 or (800) 322-8755.

You can find Ferguson on the World Wide Web at http://www.fergpubco.com

Produced by Bright Futures Press (http://www.brightfuturespress.com)
Series created by Diane Lindsey Reeves
Interior design by Tom Carling, carlingdesign.com
Cover design by Salvatore Luongo

Table of Contents Frans Lanting/Corbis; Page 5 Condor36; Page 7 Paul Colangelo/Corbis; Page 10 Judy Marie Stephanian; Page 13 Glen Jones; Page 14 Don Hammond/Design Pics/Corbis; Page 16 Christopher Vika; Page 18 Tan Kian Khoon; Page 21 Scott David Patterson; Page 22 John Sartun; Page 25 Kippy Lanker; Page 26 Dale A. Stork; Page 29 Zuma Press/ZUMA/Corbis; Page 31 David Woods/Corbis; Page 32 Patricia Marks; Page 37 Dale A. Stork; Page 39 Scott Lituchy/Star Gazette/Corbis; Page 42 William Attard Mc-Carthy; Page 40 Tony Navarro; Page 43 Emily Lo; Page 44 Bob Grinstead; Page 47-54 Cheryl Gottschall.

Note to Readers: Please note that every effort was made to include accurate Web site addresses for kid-friendly resources listed throughout this book. However, Web site content and addresses change often and the author and publisher of this book cannot be held accountable for any inappropriate material that may appear on these Web sites. In the interest of keeping your on-line exploration safe and appropriate, we strongly suggest that all Internet searches be conducted under the supervision of a parent or other trusted adult.

Printed in the United States of America

VB PKG 10 9 8 7 6 5 4 3 2 1

This book is printed on acid-free paper.

CONTENTS

Welcome to the World of Firefighting!

How can you tell a firefighter from the rest of the crowd at a fire (besides the big red truck and telltale uniform!)? They are the people running *into* burning buildings when everyone else is running *out*.

By any standard, this trait requires nerves of steel, impressive physical prowess, and a commitment to getting the job done—no matter what the cost. By any measure, it's not the typical way to earn a paycheck.

But ask any firefighter why they do it and you're likely to get the same response—"There's nothing I'd rather do." Many claim it's the best job in the world. And, the rest of us must agree since firefighter makes regular appearances on "most respected professions" lists.

So what is it about firefighting that draws some of the best and brightest among us? What is it that keeps so many dedicated men and women on the job day after day, year after year, in spite of the danger lurking behind every alarm? Just what is it about this job that causes nearly every kid on the planet to fantasize about growing up to be a firefighter?

Read on and find out for yourself about

• the long and proud history of firefighting

• what it's like to be in the "hero" business every day

• some of the 21st century trends and technology that are changing the way firefighters do their jobs

• what it takes to make it as firefighter

• what real firefighters say when kids like you ask what it's *really* like to be one

Then, get ready for the ultimate virtual firefighting experience and be a firefighter for a day. And as Assistant Chief Michael L. Painter of the Missoula, Montana Fire Department says every chance he gets, "Stay low!"

Getting the wet stuff on the red stuff is what firefighting is all about.

Smoke, Fire, and a Whole Lot More

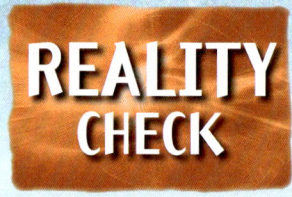

REALITY CHECK

What would you do if you found yourself in an emergency situation?

A Call 9-1-1 and stick around to help?

B Call 9-1-1 and run for the hills?

Imagine this…You're sound asleep in a humble village of yesteryear only to be awakened by the clanging of church bells and the sound of the night crier's panicked voice yelling "Fire!" Like other members of your family, you jump up, throw on some clothes, stumble into your shoes, and run out the door. On your way to the street, you grab one of the wooden buckets kept for this very purpose on the front porch and run toward the village well. Along the way you pass other neighbors, buckets in hand, rushing toward Main Street where smoke is rising and flames are shooting toward the sky. At the well you join your neighbors in one of the earliest forms of firefighting–the bucket brigade.

Two lines of people work feverishly to extinguish the fire by passing buckets of water up and down two lines between the well (the nearest source of water) and the blacksmith's shop (the source of the fire). When the smoke has faded, it is clear to see that, even though the shop is in ruins, the fire has been quenched and the rest of the town has been saved.

Hurray for the bucket brigade! It may seem simple but this firefighting system was the only option available back in the 1700s when this story takes place.

Firefighter: A person who runs into burning buildings when everyone else is running out.

Fast forward a couple centuries…

It's tomorrow night. Your family has been asleep for hours when you are jolted awake by the blaring noise of a smoke alarm. You hear your mom running down the hall, calling your name, and dialing 9-1-1 on the cordless phone. Your dad has your little sister in one arm as he scoops up the family pet with the other. You all rush outside to the neighbor's house and in a matter of

Firefighters were among the first responders when terrorists attacked the World Trade Center on September 11, 2001.

minutes a big red fire truck, with sirens blaring, comes roaring down your street. While emergency medical technicians (EMTs) check everyone for smoke inhalation and shock, a crew of well-trained firefighters attacks the fire with high-pressure hoses, ladders, and other equipment. The fire is contained quickly and, to everyone's great relief, the damage is minimal. Fire inspectors rush in to determine the cause of the fire and to make sure there are no lingering problems waiting to erupt.

Needless to say, firefighting has come a long way since the days of the bucket brigade. In some ways, fire is less of a threat. In other ways, the stakes are higher than ever. While everyday necessities like candles used for light and fire stoves used for heat and cooking were common fire hazards in the past, today's homes possess many safety features that make house fires less commonplace. However, even fiercer fire hazards have emerged in ways our ancestors never dreamed possible.

Fire!

Fuego! Fuoco! Fire! No matter what language you say it in, fire is scary stuff. So what is fire, anyway? Where does it come from? And why does it scare the wits out of people everywhere? The Merriam-Webster dictionary defines fire as "the phenomenon of combustion manifested in light, flame, and heat." Wikipedia, the free online encyclopedia, describes it as "the combination of the brilliant glow and large amount of heat released during a rapid, self-sustaining exothermic oxidation process of combustible gases ejected from a fuel." Huh?

Maybe it will help to know that fire is made up of three ingredients: fuel, oxygen, and heat. Alone these ingredients are harmless and actually very helpful. For instance, we'd be goners without a steady supply of oxygen. The potential for trouble comes when the three get together and–kaboom!–combustion happens. When not carefully controlled, the situation can get out of hand real quick.

No matter how a fire starts, whether it's big or small, indoors or out, a firefighter's main priority is, as they often say, to get the wet stuff on the red stuff. There are basically three ways to stop a fire: Take away the fuel, take away the oxygen, or take away the heat. That, in a nutshell, is what being a firefighter is all about.

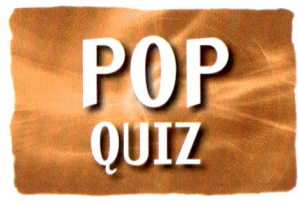

POP
QUIZ

Who Said It?

"Only you can prevent wildfires."

ANSWER: Smokey Bear, undoubtedly the world's cutest and most famous park ranger, has been issuing this warning since the 1940's. Get acquainted with him at his Web site at http://www.smokeybear.com.

Lessons Learned Along the Way

Fire has been both friend and foe to civilizations down through the ages. For centuries, it has kept people warm, provided a way to cook food, and been an essential source of energy and light. From the earliest times, fire allowed people to forge metal tools, bake pottery, and harden bricks—important linchpins of civilization. When properly contained, fire is one of the human race's greatest assets.

But when fire runs amok—watch out! Entire cities have been destroyed by raging fires—some more than once. Boston, Chicago, Baltimore, and San Francisco are all great American cities once brought to smoldering ruin by raging fires. Of course, you'd be hard pressed to find much evidence of these tragic events now—except for one thing. Some of the fire safety regulations you are likely to encounter every day are the direct result of lessons learned from these dreadful fires.

For instance, in 1903, after a horrible fire in the popular Chicago Iroquois Theater left 602 people dead, officials examined the causes and discovered many ways to make public buildings safer. Many of the fire codes and regulations we have today are a direct result of tragic fires like this one. Good things like limits on how many people can be in a building at one time, fire alarms, and even the fire engines we use today have all come about because of what firefighters learned from these bad situations.

Schools like yours are safer now because of lessons learned the hard way when fire erupted in places like Collinwood, Ohio, at the Lakeview Grammar School. In 1908, a fire raced up the main stairway of the school, trapping students and teachers who had no other way to escape. Sadly, 174 students and teachers died. As a direct result of that awful blaze, exit drills in schools became mandatory, fire departments did a better job of in-

Did You Know...

▸ Thirty seconds is all it takes for a small flame to turn into a big fire?

▸ In just a couple minutes an entire house can fill with thick black smoke?

▸ In five minutes a room can get so hot that everything in it ignites at once?

▸ A room on fire can be 100 degrees at floor level but 600 degrees at eye level?

▸ Smoke and toxic gases kill more people than flames do?

Source: U.S. Fire Administration

specting schools to look for potential fire hazards, and school officials everywhere started paying more attention to fire safety. Alarms, first-aid equipment, and additional exit routes made sudden appearances in schools across the country.

Even things that now seem obvious were learned the hard way. For instance, did you know that it's easier for people to exit a crowded building if the exit doors swing out with the flow of traffic? No one had thought about this simple strategy until hundreds of people were trapped in a raging fire in 1942 at Boston's Cocoanut Grove nightclub. No one could get out because they were jammed against the exit door. Four-hundred and ninety-one people died that night, and some experts believe that as many as 200 might have been saved if the exit door simply swung open the other way.

Dalmatians are traditional fire station mascots.

Twenty-first Century Firefighting

Despite all the valuable lessons learned through the years, fire continues to be a formidable enemy in modern times. To get an idea of what U.S. firefighters are up against, take a look at some of these statistics. According to the National Fire Protection Association, in 2004, for instance, firefighters across America responded to 1,550,500 fires. These fires killed 3,900 civilians and 103 firefighters and injured 17,885 people. The cost to repair damages caused by these fires ran a steep $9,794,000,000 (yes, all those zeros are correct!). Add what the National Interagency Fire Center reports as another 77,534 wildfires devouring some 6,790,692 acres of forest and woodlands and you've got an awful lot of fire-induced mayhem going on.

Of course, fires aren't the only catastrophes firefighters deal with. Today's firefighters have to be ready for anything and are trained to respond to everything from automobile accidents to plane crashes, health emergencies to hazardous materials spills. Responding to any kind of natural disaster Mother Nature might throw our way is part of the job too—whether it's a tornado, hurricane, blizzard, or flood. And, as the world sadly learned on September 11, 2001, firefighters are sometimes called upon for help in the awful aftermath of terrorist acts as well.

Firefighting Weapon of Choice

As exciting as putting out fires may be, firefighters would still rather prevent them than fight them. Which is why firefighters everywhere spend a lot of time and attention on prevention efforts. And since the vast majority of fires are the direct result of the human "oops" factor, firefighters go directly to the source to get the word out to people young and old at schools and other community events.

All these efforts pay off in ways that save lives and, as its name implies, prevent fires. Is there a kid in America who doesn't know the famous firefighting advice to "stop, drop, and roll" if their clothes ever catch fire? And, isn't "9-1-1" one of the first phone numbers most people memorize as children? You get the point, don't you? Fire prevention works.

A Firefighting Future

They fight our battles with fire. They rescue us. They teach us how to stay safe. No wonder people of all ages admire firefighters so much. In fact, according to a Harris Poll conducted in August 2004, firefighting ranked as one of the most prestigious professions of all (only scientist and doctor scored higher). But, not only do other people like what firefighters do, firefighters like what they do too. Job satisfaction runs high among firefighters everywhere. They simply like what they do for a living.

What about you? Can you imagine yourself one day standing nose to nose with a raging fire? Do you see yourself dashing into a burning building to rescue someone stuck inside? By the time you finish this book, you may decide that's exactly what you want to do when you grow up. Or you may decide that you'd rather watch in fascination as firefighters do their heroic work. Either way, you're bound to find that firefighters deserve every bit of respect that they get.

FUN FACTOID

BAD NEWS: There were 1,550,500 fires in the U.S. in 2004.

GOOD NEWS: That's 415,000 fewer fires than there were in 1995!

Source: U.S. Fire Administration

CHAPTER 2

Fighting Fires, Saving Lives, Doing Chores

Admit it. When you think firefighting you think of brave people dashing into burning buildings, daring rescues, and nonstop drama. Most people do—thanks to the wonders of modern media blasting riveting fire scene images on television and in headline news. These images are totally true. Firefighters are often called upon to respond in these heroic and noble ways. What most people don't realize is that these types of adrenaline-pumping situations are not necessarily reflective of a "typical day on the job" for most firefighters.

In fact, there's really no such thing as "typical" when it comes to describing a firefighter's job. Some days are filled with more emergencies than most people see in a lifetime while others are spent waiting (and waiting and waiting) for the big alarm to sound. But waiting doesn't mean that firefighters sit around twiddling their thumbs. Not by a long shot. Every day is full of such wide and varied tasks that most firefighters report that no two days are alike.

One of the most important things firefighters do after an emergency call is get ready for the next one. Fire trucks must be cleaned and kept in tip-top shape. Every piece of

FUN FACTOID

Benjamin Franklin founded the first American volunteer fire company in Philadelphia in 1736.

"Firefighters – your worst nightmare is just another day at the office."

—AUTHOR UNKNOWN

equipment must be returned to good-as-new condition. So whether each tool has to be cleaned, repaired, or replaced, everyone does whatever it takes to make sure everything is ready when the next alarm rings. Supplies must be replenished. Air tanks refilled. Uniforms, boots, and helmets readied for firefighters to jump back into them at a moment's notice.

Ready and waiting for the alarm to sound.

"Firefighter - one of the **few professions** left that still makes **house calls.**"

—AUTHOR UNKNOWN

Firefighters are protected from head to toe with high-tech gear.

Home, Sweet Firehouse

Working at a fire station is different from working in an office in lots of ways. One of the biggest differences is that firefighters not only work at the station but they live there, too. Instead of working eight-hour shifts on Monday through Friday, like most people do, firefighters tend to work 24-hour shifts. This round the clock schedule means that firefighters work, eat, and sleep at the fire station when they are on duty. But it also means that they get time off to go home two out of every three days—a perk most firefighters enjoy.

So anything that people have to do in a home, firefighters must do at the fire station. Make no mistake, there's no maid service here. No moms to make the beds or dads to cook the food. Firefighters take turns cooking, cleaning, and shopping for food. And no bickering about whose turn it is to do what is allowed!

Practice Makes Perfect

Training is another way that firefighters spend their time. Being ready for anything involves thinking through the most drastic "what if" situations and preparing to respond. What if we get to a fire and there are three children trapped upstairs? What if a train coming through town derails and spills toxic chemicals all over Main Street? What if there's a 10-car pile-up on the highway during rush hour?

The maxim "Practice makes perfect" is never as true as it is for firefighting. With so much at stake, there's little room for error. Get sloppy or forget the rules and lives can be lost and property damaged. This is why all fire departments make training a regular part of the schedule. Some departments conduct at least one training exercise on every shift.

When the world counts on you to be ready for anything, training is not something you can take lightly. In fact, ask around at any fire station and you're likely to discover that many firefighters credit good training with saving lives, keeping them safe, and making what could be a very scary job a lot easier to handle.

Another important part of preparation is physical training. They say that working a fire hose is a lot like wrestling a pine tree. It's heavy. It doesn't bend. And it bucks. With that in mind it's easy to understand why strength, agility, and flexibility are essential, not optional, in this line of work.

Couch potatoes beware! Working out and staying in shape are part of a day's work for firefighters everywhere. You'll know that you've got what it takes when you can toss a 200-pound person over your shoulder and carry him down a ladder!

Prevention Is the Name of the Game

Prevention activities are another big part of a firefighter's day. Whether it's hosting a group of school children who are visiting the station or dropping by a classroom to talk with students, many firefighters make teaching kids about fire safety a top priority. Other times they might show a homeowner how to use

Cooking Up a Fire

POP
QUIZ

Fire needs three things to survive. Use the following clues to figure out the three ingredients included in what scientists call the "fire triangle."

1 Your body needs it, cars need it, and so do fires. "It" starts with "f" and ends in an "l."

2 It's invisible, it's invaluable, and it's known to the more scientifically inclined by its symbol "O."

3 If you can't stand the _____, get out of the kitchen.

Then go online to find out more at http://www.smokeybear.com/elements.asp.

ANSWERS: 1. Fuel, 2. Oxygen, 3. Heat

Firefighters do a lot more than put out fires.

a smoke detector or help a parent install their new baby's car seat properly. Since firefighters see firsthand what happens when people aren't informed or prepared they are especially devoted to these safety and prevention efforts.

As important and necessary as all these tasks are, when the alarm sounds, firefighters stop whatever they're doing and go to the rescue. There's no telling what kinds of emergencies a given day will bring. Most days in most fire houses, firefighters are more likely to be called on for help with a medical situation or an automobile accident than they are to put out a fire. On days when fire is on the agenda, all their training, preparation, and skill comes into play—without a minute to spare.

A Company for Every Fire, A Person for Every Job

Before firefighters climb aboard a truck heading to a fire, they already know exactly what they are supposed to do. Each firefighter is assigned a specific job to perform as part of a firefighting team or *company*. Depending on their skills and training level, firefighters might be assigned to one of three types of firefighting companies: a rescue company, a ladder company, or an engine company.

Bet you can guess what a *rescue company* handles, can't you? Yes, they handle the rescue operations and are likely to be staffed by firefighters with emergency medical technician expertise. They are the ones who go inside buildings to find and retrieve anyone trapped inside.

It's best if firefighters assigned to *ladder companies* conquer their fear of heights before they get to work because they are likely to encounter a ladder or two before the fire is out. Their job is to use both aerial and ground ladders to open windows, roofs, and walls, using a variety of forcible entry techniques so that other companies can get up close and personal with the fire.

Those firefighters assigned to an *engine company* are the ones who actually get the wet stuff on the red stuff. They man the hos-

In Case of Fire...

Quick! Grab a blank sheet of paper and see if you can complete the following statements.

1 There's a fire hydrant located about _____ blocks from my house.

2 The fire station nearest my house is located at _____ _____.

3 Three ways to get out of my house if it catches fire are :

_____,

_____,

and _____.

4 It would take firefighters about _____ minutes to get from the nearest fire station to my house.

5 The fire extinguisher in my house is kept _____.

6 There are smoke detectors _____ and _____ in my house.

What? You don't have a fire extinguisher or smoke alarms in your house? Talk with your parents about getting them soon. Your local fire department may be able to help you find free or inexpensive ones.

POP QUIZ

es, secure water access, and actually put out the fire. Firefighters in engine companies are each assigned a specific position on the fire line and they all depend on each other to be where they are supposed to be and doing what they are supposed to be doing.

Here are some of the specific firefighting positions:

The **chauffeur** is the person who drives the fire truck. Not only is this firefighter responsible for safely and quickly transporting the fire crew and equipment to precise locations, but other duties kick into gear once the company reaches its destinations.

The **chief** is likely to be found wearing a white hat, calling the shots, and making sure that the job gets done and that all the firefighters are accounted for at all times. He or she is the "boss" of

Firefighters work in teams and depend on each other to get the job done.

the situation and is an expert at handling emergencies.

The **incident commander** helps size up each situation and determines how factors like weather, population, the type of building on fire, the number of nearby fire hydrants, and the location of nearby buildings might affect the firefighting effort. This person often wears earmuffs with an intercom attached so that he or she can communicate with other people back at the department and on the ground.

The **nozzle man** is the first firefighter in line at the front end of the hose line. His or her job is to direct the water stream and control the nozzle. A person right behind the nozzle man helps hold and move the heavy hose. This is called "heeling the line."

The **roof man** is a firefighter who rides up in a bucket attached to an aerial ladder to cut holes in roofs to release dangerous gases and provide a direct route for water to be applied to flames.

The **vent man** is a firefighter who uses an ax and other tools to open doors and break windows to give smoke and heat somewhere to go.

Move It or Lose It!

If you ever find yourself inside a burning building, follow these tips to get out alive.

1 Get out quick!

2 Think twice and figure out two ways to escape!

3 Stay low!

4 Keep out!

5 Call for help!

Let Sparky the fire dog help you plan your escape from fire danger at http://www.nfpa.org/sparky/family.html.

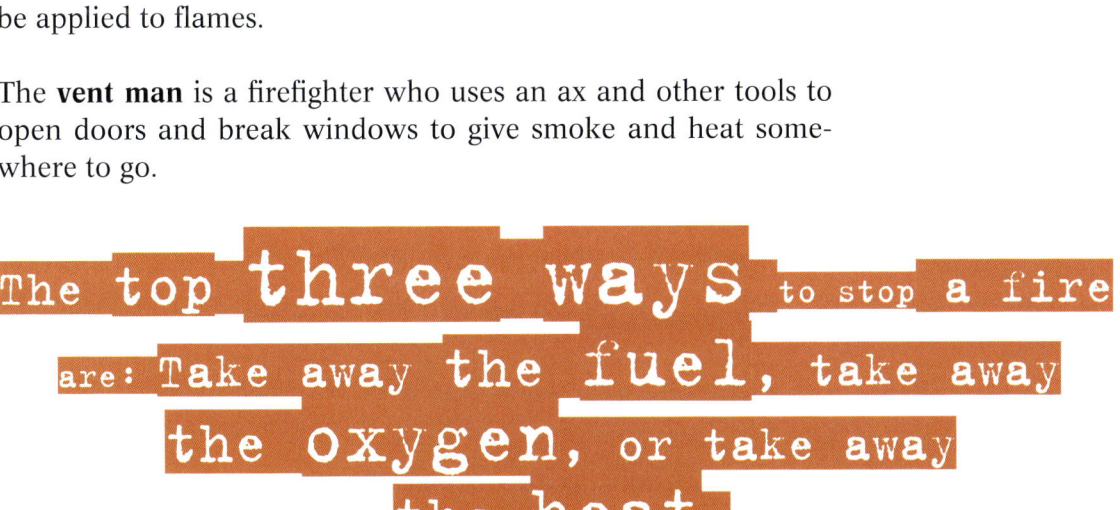

The top **three ways** to stop a fire are: Take away the **fuel**, take away the **oxygen**, or take away the **heat**.

CHAPTER 3

Firefighting Tech and Trends

When it comes to modern firefighting equipment, it's technology to the rescue. Sometimes simple, sometimes mind-boggling in complexity, technology has provided tools that make firefighting safer and more efficient than ever before.

Of course, sometimes even the best technology can't beat old-fashioned ingenuity. For instance, science has yet to come up with a better tool than an ax to chop away troublesome obstacles. And, is there an invention known to man that can jimmy open a locked door quicker than a crowbar? Firefighters actually use a halligan bar, a jazzed up crowbar with a cutting blade to increase its usefulness in prying or cutting away obstacles. But, still, you get the idea, don't you?

Beyond these basic, yet essential tools, is an impressive array of equipment that would have nineteenth-century firefighters wide-eyed with wonder. And what better place to start with life-saving gear than with the protective clothing firefighters wear to do battle with fire? After all, standing in the middle of a hot fire, the last thing you want to think about is whether or not your clothes will keep you cool and dry.

FIND OUT MORE

How does fire work? Smoke out some cool facts about fire online at http://science .howstuffwork .com/fire.htm.

Join in the fight as firefighters battle blazes at http:// express.howstuff works.com/battling-blaze.htm.

The Well-Dressed Firefighter

Firefighters have their own brand of "dress for success." In their case, success is spelled s-a-f-e-t-y and everything they wear, from head to toe, is specially designed to protect their bodies in the heat of even the hottest fires. Even their underwear—

Turn-out coats and pants make it easy for firefighters to get dressed on the run.

Smoke can be as deadly as flames, so firefighters take precautions when inside burning buildings.

all-cotton, if you please—is designed to protect them during fires and keep them cool during hot weather.

Someone or, more likely, lots of some-ones, have spent a lot of time thinking of ways to make it easier for firefighters to get into nearly 60 pounds of protective gear as quickly as possible. One brilliantly simple solution: Turn-out pants, where the pants and boots are attached in such a way that the firefighter can jump in the boots, pull up the pants, and get on the road.

And, speaking of boots, whoever designed those worn by most firefighters thought of everything. Firefighter's boots feature reinforced steel plates to keep out nails and broken glass. Nonskid treads keep them on their feet (as opposed to flat on their backs) in slippery situations. Fire-resistant rubber keeps them cool and shielded from fiery flames.

Like turn-out pants, turn-out coats share similar features that make the garment easy to slip into quickly. Every inch is carefully designed for protection—from the high collars to the extra big pockets. The coats are constructed with three layers of fire-resistant fibers. The outer layer helps protect firefighters against temperatures up to 1,200 degrees Fahrenheit. Its flame-

Firefighters wear nearly 60 pounds of protective gear to keep them safe from fire and smoke hazards.

resistant fabric was scientifically-formulated so that instead of catching on fire or melting when exposed to flame it forms a harmless char.

The second layer provides a moisture barrier, which is as much for the firefighter's safety as it is for his or her comfort. It's not just annoying to get wet. It's dangerous too. If water soaks through to the skin it can heat up, turn to steam, and cause serious burns. Another example of scientific ingenuity, the synthetic fabric used in this layer is woven with teeny tiny openings that let air in but keep water out.

The third layer is made of special synthetic fibers that trap air near the firefighter's body and cool the heat. Suspenders, snaps, and Velcro make it easier to fasten things in a hurry.

Next comes the helmet, which not only provides protection from falling debris but also offers a wide brim and inside padding to cover the firefighter's hair, ears, and neck. It, too, has a distinctive design with an especially long back. Can you guess why? It keeps water from flowing down a firefighter's neck into her or his coat.

Last on are heavy-duty, fire-resistant, and very flexible gloves.

Since exposed skin is like a magnet for fire, the goal of all this gear is to keep every inch of the firefighter's body completely covered. Like Superman in his cape and tights, once attired in this protective gear, modern firefighters are ready to take on the world. Raging fire? Bring it on!

Keep 'Em Alive

Now dressed head to toe in protective gear, firefighters are almost ready to tackle the worst fire has to offer. But not before clipping an **electronic motion alarm** onto their belts. This little device is equipped with an electronic motion sensor and timer. If the firefighter doesn't move for 30 seconds (as in, "Help! I'm down!"), the de-

Famous Fires

Chicago

Who was Mrs. O'Leary and what did her cow have to do with the Great Chicago fire? Find out at your local library or online at http://www.chicagohs.org/fire/oleary and http://www.nationalgeographic.com/ng kids/9809/chicago.

San Francisco

What was it like to be a firefighter in 1906 during the San Francisco earthquake and fire? Look for books at your local library or go online to http://www.sfmuseum.org/conflag/06index.html and read firsthand accounts from firefighters who were there when it happened.

CHECK IT OUT

vice will emit a high-pitched whistle to alert others that a fellow firefighter may be in danger.

Firefighters also strap on a **Self-Contained Breathing Apparatus** (SCBA), a 30-pound metal tank containing a 30-minute supply of compressed air. The system is equipped with a mask that fits over the firefighter's mouth, nose, and eyes. A glow-in-the-dark pressure gauge and sensor alarm alert firefighters when it's time to go outside for a fill-up of fresh air.

Most people don't realize that smoke can be even more dangerous than flames when fighting fires. Among other things, it diminishes the fresh air supply, making it difficult, if not impossible, to breathe in a smoke-filled room. Even though wearing SCBA gear makes firefighters look like scary aliens from outer space, the tanks keep them alive by providing a steady flow of oxygen.

Thermal imaging cameras are another example of recent technology that addresses a dangerous problem caused by the blinding effects of smoke. Within minutes, smoke can so completely fill a room that it's impossible to see; a situation that, needless to say, can hamper even the most courageous firefighting efforts. Thermal image devices allow firefighters to see in the dark with technology that seeks out heat. They use this technology to find people trapped in burning buildings, to make quick assessments of fire situations, and to detect hot spots hidden behind walls and ceilings.

Big Problems, Simple Solutions

Solving problems is the one thing all firefighting tools and technology have in common. Sometimes the solutions are as big as the problems themselves. But many times, it is surprising how far a little brain power can go in solving huge problems. For instance, for decades most fire trucks were equipped with hoses that were two inches in diameter. While certainly bigger and more powerful than an ordinary garden hose, they don't pack nearly as much punch

Time to Spare

When seconds count there's no time for goofing off or goofing up. See how much time you can shave off your morning routine. Think through all the things you have to do each morning—hygiene, dressing, breakfast—and see if you can come up with better ways to get the same things done in less time. Time yourself before and after you implement the new plan. Did your ideas work?

Firefighting is an equal opportunity profession.

as the newer five-inch hoses now widely used. Of course, advances in the cotton and high-tech synthetic materials used to construct these hoses contributed to this "bigger is better" trend and make the wider hoses easy to handle and maintain.

Home safety increased by leaps and bounds with the simple and, thanks to the efforts of fire departments everywhere, widespread installation of smoke and carbon monoxide alarms. These noisy devices detect the presence of smoke or poisonous gases in the air and alert people and animals to get out. These inexpensive and easy-to-install alarms are credited with saving countless lives.

Going My Way?

When a fire alarm sounds why do firefighters slide down poles instead of using the stairs?

POP
QUIZ

ANSWER: It's quicker (and safer) to slide down a pole than to run down a flight of stairs. The seconds they save can mean the difference between life and death for a person who needs help.

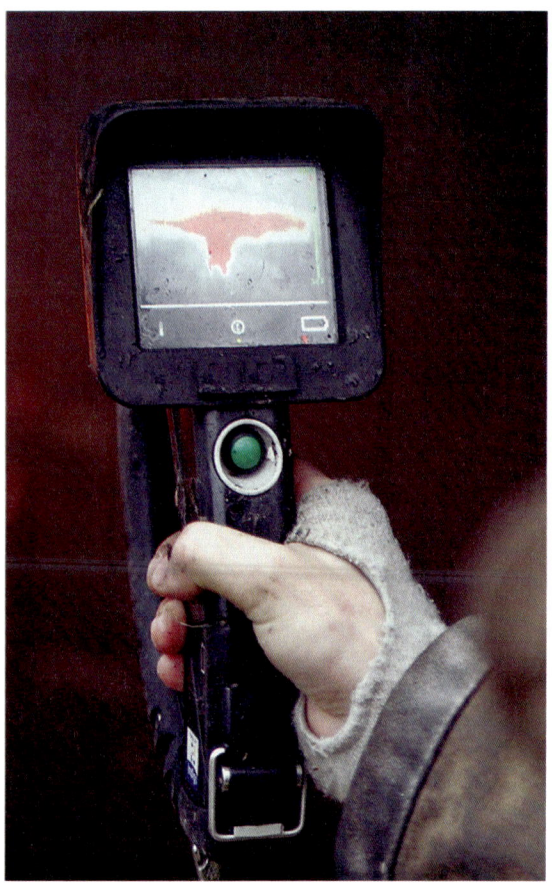

Thermal imaging cameras help firefighters see in the dark.

Fire extinguishers, another life-saving device, are now found in homes and public places everywhere. It provides a tool for responding quickly to small, contained fires. You remember that fire can be put out in any of three ways, right? Dump water on it and remove the heat, remove the source of fuel feeding the fire, or cut off the supply of oxygen. Fire extinguishers use the third method by smothering fires with a nonflammable chemical mixture that usually contains a substance very similar to baking powder or baking soda. Even though most fire extinguishers never get used (thank goodness!), they have proven themselves a worthy firefighting friend when fires do erupt in homes and in the workplace.

Firefighting in the Future

Twentieth-century technology made for some very impressive gains in equipping people to prevent and fight fire. Can you imagine what people will think of next as the twenty-first century moves along? Exciting possibilities loom ahead!

For instance, the Air Force Research Laboratory has been working to develop an ultra-high-pressure firefighting system that combines high-pressure water and a foam solution containing carbon dioxide. Two things make the new system more efficient than current methods: the interaction of the water and carbon dioxide and the water pressure. While normal fire trucks deliver water at a rate of 500 to 700 gallons per minute, fire trucks equipped with this new system deliver at a significantly lower 200 gallons per minute. In this case, less is definitely more in terms of firefighting efficiency due to the system's ultra-high-pressure technology. The bonus is that because less water is needed to douse flames, the Air Force (and eventually all community fire departments) could use smaller, lighter vehicles to transport water—a big plus in war zones and peaceful neighborhoods alike.

Some people think of firefighters as emergency workers, or 9-1-1, without guns. But all that could change if a new device currently in the works becomes commonplace and a new firefighting gun takes center stage. The idea behind this up-and-coming device is to apply what scientists know about rockets and the aerodynamics of liquid and gas mixtures to create a powerful firefighting weapon that shoots fire-dousing chemicals. The result so far is a lightweight, environmentally friendly device that doesn't need a big red truck to carry it from fire to fire. That factor alone could eventually earn this device a place in the firefighting hall of fame! This type of technology could make immediate and effective response to fires possible in places like shopping malls, apartment complexes, businesses, gas stations, mines, and other public places. Oh, and did we mention that early tests indicate that it can quench fires in a mere fraction of the time and with considerably less water than a traditional fire hose?

Put promising developments like these together with the technological advances of the past century and you can't help but be amazed at how much safer and effective firefighting has become. Will fighting fires ever be risk-free and easy? Of course not. But every life saved makes these high-tech efforts worth it.

FUN FACTOID

George Washington, Benjamin Franklin, John Hancock, Alexander Hamilton, Samuel Adams, and Paul Revere were some of early America's most famous volunteer firefighters. Find out more about their firefighting history at http://www.firefightersrealstories.com/volunteer.html.

CHAPTER 4

Brains, Brawn, and Heart

CHECK IT OUT

Jaws of life. Backdraft. Joker. Huh? Firefighters have a language all their own. Use an Internet search engine like Google (http://www.google.com) or Yahoo (http://www.yahoo.com) to run a search for "firefighter terminology." While you're surfing, see if you can figure out the difference between a fire tornado, a fire curtain, and a firewhirl.

When you were younger, you probably heard a lot of grown-ups say things like "Look both ways when you cross the street," "Don't run with scissors," and "Keep your fingers out of electrical sockets."

These adults were trying in their own well-meaning (and sometimes annoying!) ways to teach you how to survive some of life's everyday dangers. Since you are sitting there reading this, their efforts at training must have worked so far. You haven't been run over by a car. You haven't poked out your eyes with scissors. And, one can only hope, you haven't nearly French-fried yourself with an electrical shock.

If kids need all that training just to make it through an ordinary day, just think about all the training firefighters need to do their jobs!

Getting Started

Once upon a time, if you wanted to be a firefighter all you had to do was apply at the local fire station and take some tests. If you passed you went off to a firefighting academy for a couple months and, if you made it through, you were hired. Those days are long gone.

"What you call a hero, I call just doing my job."
—UNKNOWN FIREFIGHTER

Every second counts in emergency situations like this.

"We **are** what we **repeatedly do.** **Excellence** then is **not an act,** but a **habit.**" —ARISTOTLE

For one thing, as we've mentioned before, there's a lot more to firefighting than there used to be. New technology and new challenges make the job more complicated than it may have been for, say, your great-grandparents.

For another thing, lots of people want to be firefighters. It's a cool job with good pay, reasonable hours, and nice benefits. When lots of people apply for firefighting jobs, the people who get hired are the ones with the best training and experience to do the job.

So what does that mean for aspiring firefighters? Simply this—do your homework. First, find out all you can about the specific requirements for becoming a firefighter in the place you

Repeat Performance

Maybe you did it when you were a little kid. Visited the local fire station, that is. If so, you, along with virtually every other youngster in America, were probably mesmerized by the shiny red trucks, the sirens, and all that gear. Chances are you even fantasized about growing up to become a firefighter some day.

Now that you're older and wiser, schedule another visit. Ask if you can hang out for an hour or a day and see what it's really like to be a firefighter. Look around. Ask questions. View everything you see through a "Is this a good choice for me?" lens.

If the answer is a resounding yes, or even slightly more than maybe, take things a step farther and get involved. Ask around about opportunities to become a volunteer firefighter in your area. In some places, you can get started while you are in high school. Go online to the Fire Corps Web site at http://www.firecorps.org for more information.

Find out about local Fire Service Exploring Clubs (for kids!) at http://www.learningforlife.org or sign up for a Red Cross first aid or CPR course at http://www.redcross.org.

CHECK IT OUT

Firefighters say there is nothing better than rescuing a child.

want to work. In most cases, the Internet makes it easy to go online and find useful information about the hiring practices of local fire departments.

Then, depending on what they're looking for, consider pursuing one or more of the following training strategies to give yourself a competitive edge:

Get a two-year associate's degree in fire science through a local community college. These programs will provide a thorough introduction to firefighting basics and should include courses like fire protection, fundamentals of fire prevention, hazardous materials, chemistry for fire service, and fire protection systems and equipment. Hint #1: If you like the courses, you'll probably like the work and vice versa.

Get certified as an emergency medical technician (EMT) or paramedic by completing the requirements for basic certification

Firefighting is tough work.

(sometimes referred to as EMT-1). This will take six months to two years of both classroom study and hands-on clinical work and will prepare you to do things like perform physical exams, assess trauma, administer oxygen, perform airway maintenance, and administer medication. Hint #2: See hint #1!

Get experience as a volunteer firefighter and prove to yourself and others that you've got the right stuff. Since, according to the National Volunteer Fire Council, there are three volunteer firefighters for every paid one, it shouldn't be too difficult to find a way to serve. You can find out more about local volunteer fire-

"**Practice** does not make perfect. Only **perfect practice** makes perfect."

—VINCE LOMBARDI, FAMOUS FOOTBALL COACH

fighting opportunities at the Fire Council's Web site at http://www.nvfc.org.

Put Your Best Foot Forward

Getting the brain power, through training and education, is only the first step in preparing to become a firefighter. Getting the "brawn" is just as important. "Brawn" refers to physical strength. It's something you'll need in great supply since, in case you haven't noticed, firefighting is one of the most physically challenging jobs on the planet.

Transforming oneself into a lean, mean firefighting machine takes attention to three types of physical training. Start with some good stretching exercises to increase flexibility. Follow-up with some rigorous cardiovascular work like running, swimming, or aerobic exercise. Top it all off with some weight training to build strong (make that buff!) muscles. Keep this up several times for a week over the course of a 20- to 30-year career and you'll be in the kind of shape needed to fight fires.

Firefighter for Hire

After all that preparation, a prospective firefighter is ready to submit applications and start interviewing with fire departments looking for recruits. Those who make it through this screening process are subjected to a round of tests. This includes a written entrance exam to test basic aptitudes and a psychological test to determine mental and emotional fitness for the challenges of firefighting. Physical fitness tests are conducted to find out if would-be firefighters can handle some of the physi-

Little Red Schoolhouse

You won't find courses like these on the schedule at your school, but you will find them at one of the nation's most elite firefighter schools, the Emergency Services Training Institute at Texas A&M University.

- Fire Behavior
- Ropes and Knots
- Building Construction
- Sprinklers
- Search and Rescue
- Forcible Entry
- Ladders
- Fire Hose
- Extrication Tools
- Extrication Techniques
- Auto Extrication
- Ventilation
- Water Supplies
- First Aid
- CPR
- Technical Rope Rescue

FIND OUT MORE

Find out more about this amazing fire training program at http://teexweb .tamu.edu/ESTI.

cal demands of firefighting, like making an aerial ladder climb, dragging a charged hose, carrying a victim, and running through a rescue maze.

Candidates who pass all these tests are invited to attend training at a firefighting academy. That's where the real fun begins. Just like military boot camps, training at fire academies is really intense for a very good reason–to challenge the "cream of the crop" to rise to the top. This process weeds out the so-so candidates from the bunch and provides something of a guarantee that those who make it through this three- to six-month training truly have what it takes to make it as a firefighter.

Recruits at a firefighting academy go through a variety of learning experiences–both in the classroom and in realistic hands-on exercises–that fully acquaint them with firefighting techniques, fire prevention, hazardous materials control, local building codes, and emergency medical procedures, including first aid and cardiopulmonary resuscitation (CPR). There are plenty of opportunities to learn how to use axes, chain saws, fire extinguishers, ladders, and other firefighting and rescue equipment.

Those who pass these challenges with flying colors are assigned to various fire departments or companies on a probationary basis. As "probies" they join in the regular training exercises with their fellow firefighters. And they soon learn that, even for the most seasoned firefighters, there's always something new and important to learn about firefighting.

A Little Heart Goes a Long Way

There's no doubt that successful firefighters are smart and they're tough. But there's another, perhaps surprising, characteristic necessary to make it as a firefighter, and that's heart. The people firefighters regularly encounter are having one of the

Firefighting ABC's

Bet you didn't know that fires get grades, too. Firefighters rank different types of fires like this:

▶ Class A fires involve ordinary materials like wood, paper, plastic, rubber, and cloth.

▶ Class B fires involve flammable liquids or gases like gasoline, propane, and many types of oils.

▶ Class C fires involve electrical equipment like household appliances and power lines.

▶ Class D fires involve combustible metals like aluminum, magnesium, titanium, and lithium.

Source: American-Firefighter.com

worst days of their lives. They are scared, stressed-out, and, in some cases, severely injured. Some are watching the only home they've ever known go up in flames. Others are grappling with fears that someone they love has been hurt or worse.

Firefighters must be ready to deal with this human side of their job. They need to be strong, but they need to be sympathetic, too. They need to be able to communicate clearly and with the kind of confidence that inspires trust.

Caring about other people seems to come naturally to the best firefighters. But somehow they must strike the delicate balance of keeping their hearts soft with compassion and their minds tough enough to do what they have to do to get the job done.

Couch Potatoes Beware!

Firefighters have to stay in tip-top shape so they are ready for any of the physical challenges that come their way.

Take this Virtual Apprentice firefighter challenge: Do 20 push-ups, 20 sit-ups, and 20 jumping jacks without stopping for snacks, naps, or other diversions.

Ready for more? Find ideas at the Verb: It's What You Do Web site at http://www.verbnow.com.

CHAPTER 5

Up the Ladder
and Behind the Scenes

Fires are not the only places where firefighters climb ladders. They also encounter them throughout their careers as they work their way up from "probies" or probationary firefighters to various levels within the firefighting system. Fire departments are organized in ways very similar to the military. A "chain of command" is established that clearly defines who is in charge of whom and what. Every person has someone directly ahead of him or her on the chain that they receive orders from and must answer to.

At the top of the chain are **fire chiefs**, who are, in simple terms, the "bosses." Chiefs are highly experienced officers who lead and manage fire battalions and departments. Things they concern themselves with include training, safety and health, communications, fire investigation, finance and budgeting, professional development, incident command, hazmat (short for hazardous materials, such as poisonous chemicals) response, and vehicle maintenance.

Next on the chain of command are **captains**, who tend to be in charge of the day-to-day operations at a specific fire station. Captains and are responsible for everything that happens at the station from scheduling to training programs.

"I have no **ambition** in this world but **one**, and that is to be a **firefighter**..."

—CHIEF EDWARD F. CROKER, FDNY CIRCA 1910

Working side-by-side with captains are **lieutenants**, who assist with many of the administrative tasks and management duties that come with keeping a fire station running around the clock.

Fire engineers have, through a combination of experience and training, proven themselves responsible enough to handle the fire trucks and other major equipment found in fire departments. Their main job involves taking care of driving, operating, and maintaining a specific type of equipment.

Most of the people working at a fire station are called **firefighters**. However, even at this level there are various ranks that bring additional responsibility, pay, and even prestige. Every one starts out on a probationary or apprenticeship level. Then, depending on the way their department does things, they may advance to positions designated as Firefighter 1, Firefighter 2, and other levels. Length of service, training, and performance are all factors in how fast and how high a firefighter climbs to these new positions.

Dispatchers are the cool, calm, and collected professionals who answer the phone when people with emergencies dial

Fire chiefs wear white hats so it's easy to tell who is in charge.

NAME: # Captain Gilbert Cook
OFFICIAL TITLE: # Fire Department Liaison Officer

What Do You Do?

I help coordinate local and federal rescue efforts at the Pentagon. My job was created after the 9/11 terrorist attack so that we would be better prepared to respond to major fires and threats involving weapons of mass destruction. The job is unique in that it integrates the communication and response plans of every agency likely to be involved in a major emergency situation.

I was chosen for this job because of my experience in four areas: emergency medical services (EMS), firefighting, hazmat (hazardous materials), and technical rescue. Even though I have no military experience I work at the Pentagon, which is headquarters for all of the nation's military branches.

I work with other officials to plan for all kinds of "worst- case" situations. So far we've involved police, fire, and EMS people in eight different kinds of planning exercises where we respond as if something bad has happened and practice what to do about it.

Why Do You Do What You Do?

Firefighting does seem like an unusual choice for someone who went to college to study business administration. I went to a school in the Shenandoah Valley where there were lots of opportunities to fish and hunt. At one point, a friend of mine was badly injured and I didn't know what to do to help. That helpless feeling led me to take a class in first aid and then in CPR.

I ended up coming back to northern Virginia where I studied to become a paramedic at George Washington University. I started out volunteering as a paramedic and that led to a job with the fire department. One thing led to another until I was ready to take on the challenges of my current position.

ON THE JOB

My job is to think through all the 'what ifs'.

Only qualified people fight fires. It doesn't matter if they are male or female, black, white, Hispanic, or of Asian descent.

"9-1-1." They are the link between the scene of the emergency and the source of help. As such, they must be able to pay attention to details, get their facts straight, and communicate urgent needs accurately and quickly.

A Daring Breed of Firefighters

As if fighting fires weren't exciting (and dangerous!) enough, some types of firefighters work in even more challenging situations. For instance, **smoke jumpers** are highly trained firefighters who parachute into remote locations to fight forest fires. They tend to work seasonally during the summer and fall months when warm

"9-1-1 operator. What is your emergency?"

NAME: **Tony Navarro**

OFFICIAL TITLE: **Smokejumper**

What Do You Do?

I jump out of airplanes in wilderness areas to put out fires. My job is to get to wildfires before they get too big. When a wildfire is spotted, I grab my personal gear (which is always packed and ready) and jump into a plane with about 18 other people. Once we get to the area and the cargo boxes containing our equipment have been dropped in, we get to work digging a trail around the fire as quickly as possible. This helps the fire die down. After the fire cools down we do what we call "potato patching" to chop up the ground around the fire area to get the air out and mix up the dirt. We also use chain saws to cut out trees and brush that might keep the fire going.

All this can take two or three days, so we have to make sure that we have enough food and water to last until we get the fire put out. It's like a big camp-out.

Why Do You Do What You Do?

I got started working on a hotshot crew when I was going to college in Oklahoma. One day one of my forestry professors came up and told me the fire service was looking for people to help out during the summer wildfire season in Wyoming so I signed up. Hotshots, by the way, do the same type of work as smokejumpers except that they go in by busses or helicopters instead of jumping out of planes.

A couple of years later I met some smokejumpers and got started getting the training I needed to work with them. Now I work as a smokejumper in the summer months and teach high school in Montana the rest of the year.

As a **smokejumper**, Navarro is trained to jump out of airplanes to help put out wildfires.

temperatures and dry weather create conditions favorable to fire in forested areas.

Wildland firefighters work in or near national parks or forests. Their jobs often require them to be part firefighter and part lumberjack, since eliminating the source of fuel for forest fires–trees–means chopping them down. Sometimes these types of firefighters work on what are called "hotshot" crews. By all accounts hotshot crews are elite firefighters who battle some of the biggest and meanest blazes in hot, dry natural areas.

Industrial firefighters work for companies with large, potentially dangerous, facilities like factories, mines, or airports. The wide variety of hazardous materials typically used in places like these make this particular type of firefighting even more risky.

NAME: **Robert Gillen**

OFFICIAL TITLE: **Retired NYFD Firefighter**

What do you do?

For over 20 years I was a firefighter assigned to Engine Company 228 in Brooklyn, New York. Now that I'm retired, I volunteer at the New York City Fire Museum (http://www.nycfiremuseum.org). The reason I can't seem to stay away is that I never met a firefighter who didn't love their job.

I am no exception. My job was always exciting, never boring, and even with 26 years under my belt, no two days were ever the same. Which should come as no surprise, since I spent most of that time in front of the action as the nozzle man.

Why Do You Do What You Do?

Like many firefighters, I'm not the first one in my family. My stepfather was fight-ing fires in New York back in the 60s and 70s. One thing I noticed about firefight-ing that has changed since then is that it has gotten a lot safer. I think smoke alarms, sprinklers, better construction, lots of training, and people making a lot of noise about fire safety made my job a lot easier than it was for my stepdad.

ON THE JOB

One of the things I like best is that being a firefighter automatically admits you to an elite, close-knit club. A firefighter can go anywhere in the world and be welcome at a fire station. It's what I miss most about being a full-time firefighter—the camaraderie, the teamwork, the firefight-ing family.

Firefighting aircraft are used to fight forest fires.

Behind Every Good Firefighter

Firefighters are free to concentrate on what they do best—fight fires—because there are lots of people working behind the scenes to help them. For instance, depending on the state where they work, **fire marshals** are responsible for fire safety code adoption and enforcement, fire and arson investigation, fire incident data reporting and analysis, public education, and advising public officials on fire protection. In some states, says the National Association of State Fire Marshals, fire marshals are responsible for firefighter training, hazardous materials incident responses, wildland fires, and the regulation of natural gas and other pipelines.

Stopping fires before they get a chance to get started is what **fire inspectors** are all about. They make regular inspections of public places and new construction sites to make sure that everything that can possibly be done to prevent fires is done.

NAME: Captain Emily Lo

OFFICIAL TITLE: Captain

What Do You Do?

As captain of Davis Engine #31, my number one priority every day is to make sure that all of my firefighters make it home safely each day. Firefighting is a risky business and one that you can't do alone. Teamwork, training, and looking out for each other are ways I make sure my station is prepared for anything and why I especially enjoy working at my home away from home.

I often come to work at the Davis Fire Department in California wondering how many people I'll get to help that day. Even though chances are slim—only about five percent—that there will be a fire to fight on any given day, there's no end to the ways we help the public. No two days are alike. Traffic accidents, hazardous materials, floods, and other kinds of emergencies keep my station hopping 'round the clock.

Why Do You Do What You Do?

I am one of a growing number of women working in the firefighting ranks—even though when I was young it never occurred to me that I could become a firefighter. When it came time to graduate from high school, I found myself at the "What am I going to do now?" crossroads that everyone experiences. A friend of mine had already graduated and gone on to the firefighting academy. The stories he told about his experience there encouraged me to give it a shot. I figured that I had nothing to lose and everything to gain. Little did I know that attending the fire academy would change my life.

I recall that everything about it just clicked and I realized that this was a career that challenged me physically and mentally. It was a career that would let me not only help others but also help me become self-sufficient and secure. And, it was a career jam-packed with the excitement of the unknown.

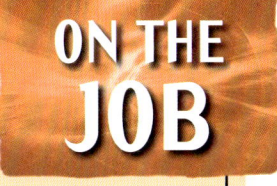

ON THE JOB

NAME: **Bob Grinstead**

OFFICIAL TITLE: **Project Engineer**

What Do You Do?

I design new vehicles for a company that is famous for manufacturing fire trucks. My work is kind of like building something with Legos. I start with a blank slate. Then I talk with customers about problems and look at ways to make things work better. Every idea is like adding another Lego to the structure.

For instance, one of the products I worked on was the Gladiator Evolution, a new state-of-the art fire truck. One of the things we looked at was how to make it more comfortable for firefighters. We found out things that only a firefighter would know. Like how they would appreciate seats with more

space in the hip area so their belts and SCBA tanks fit in better.

Why Do You Do What You Do?

I'm actually an artist by nature and think that my creativity really helps me do my job. I started working for Spartan when I was still in high school as part of a co-op program where I went to classes for half the day and worked for them the other half. I started out in the drafting department making drawings and they liked my work. That was 11 years ago, and I'm still here. Everything I know about engineering I learned on the job.

You can take a look at the kinds of trucks we design at our company Web site at http://www.spartanchassis.com.

ON THE JOB

I'm actually an artist by nature and think that my creativity really helps me do my job.

There are also all kinds of **fire safety engineers and researchers** who design and develop the tools, equipment, and technology used in the firefighting system. Everything from the protective clothes firefighters wear to the gear they use and the vehicles they drive are imagined by these creative types.

And, don't forget the **fire equipment manufacturers** who actually take the good ideas that engineers and researchers come up with and turn them into actual products that can be used in fire stations.

The Who, What, and Where of Firefighting

Who fights fires? Qualified people. It doesn't matter if they are male or female, black, white, Hispanic, or of Asian descent. As long as they've proven they've got what it takes to do a good job, anyone and everyone is welcome and encouraged to apply for these very popular jobs.

Currently there are more men than women who work as firefighters, but that doesn't mean there isn't plenty of opportunity for women. According to the Women in Fire Service Web site (http://www.wfsi.org), there are 6,200 women who work as full-time, paid firefighters. Of those several hundred have risen to the rank of lieutenant or captain and another 150 are district chiefs, battalion chiefs, or division chiefs. Another 35,000 to 40,000 women serve as volunteer firefighters.

Where do firefighters work? They work in cities big and small, in rural communities, for the federal government, for companies with high fire protection needs like hotels and amusement parks, and other places where large groups of people and valuable property and natural resources need extra protection.

What is firefighting all about? Sure, it's about fires. But it's also about protecting all that society holds dear–its people, its homes and businesses, and its way of life. No wonder firefighters like what they do so much!

CHAPTER 6

Kids Ask, Firefighters Answer

If you're like the young journalists at the Moore Square Museum Magnet School in Raleigh, North Carolina, you've got your own questions about firefighting. Maybe they are the same questions posed by this talented and very curious team of kids—Haley G., Lauren H., Joshua R., Divya T., Josh C., Jacob T., and their advisor Milly Hodges Lester who started a school newspaper called the *M2M3 Gazette*.

For questions like these, we go straight to the source for answers. In this case, we asked firefighters associated with the Missoula Fire Department in Montana to answer the student's questions. What follows are answers from Chief Michael Painter and Firefighter First Class Bill Bennett, Firefighter Ron Brunell, Firefighter Tavis Campbell, Firefighter Brett Cunniff, Fire Captain Kip Knapstead, Firefighter First Class Jeff Knoll, and Battalion Chief Todd Scott.

What inspired you to become a firefighter?

—Lauren H.

> **Firefighter Kroll:** I wanted to become a firefighter because of the wide range of job duties it entails. Some days we may be saving someone from an icy river and other days we may be helping get a cat out of a tree.

"You never know what a day is going to have in store for you."

—FIREFIGHTER JEFF KROLL

Chief Painter: I actually tried out for the fire department at the recommendation of my stepfather, who worked for the city. I had worked at several jobs while I was going to college (I moved freight in a warehouse, helped install carpet, and worked at a retail store) and while I enjoyed them all, they didn't really seem too meaningful. I am so grateful to my new dad for suggesting that I try firefighting.

Newspaper staff at Moore Square Museum Magnet School in Raleigh, North Carolina.

Lauren H.

Firefighter Campbell: When I was in eighth grade, my family took a vacation to Yellowstone. This just so happened to be when the big 1988 fires occurred. While touring through the park, we saw many fires and I became intrigued with the difference the hard-working firefighters were making.

Firefighter Bennett: This answer is an easy one for me. I am a second-generation firefighter. My father was a firefighter, retiring after more than 30 years in the fire service. I grew up around a fire station, so to speak. I have always been interested in this job, despite trying my hand at other occupations throughout high school and college. My college degree is in forestry, but deep down I knew what I really wanted to do.

Firefighter Brunell: There are so many things that inspired me to become a firefighter, but the thing that inspired me most was watching my dad work as a volunteer fireman when I was growing up.

How did you become a firefighter?

—Divya T.

Firefighter Kroll: I became a firefighter through a broad range of classes and lots of studying. The classes ranged from medical training on one end to how to climb a ladder properly on the other end.

Firefighter Cunniff: I became a firefighter after I joined the Air Force. I went through a four-month fire academy in Texas where the entire military trains all of its firefighters. Then I worked on a military base as a firefighter in both Alaska and Kuwait.

Firefighter Bennett: In preparation for the exams, I studied and reviewed multiple subjects such as math, English, and physics, and I got myself into excellent shape. I tested for the first time in 1995 and did not make the cut high enough to gain employment. After that disappointment, I realized that I needed to improve myself significantly. In 1997 I tested again, scored very well, and was hired in April of 1998. Thank goodness for all that preparation!

Firefighter Brunell: I started out by working for the Montana State Lands wildland fire crew. After five years of working on wildland crews, I decided that I wanted to become a professional city firefighter. To accomplish this, I attended a fire college for two years. Then I began testing at various fire departments across the Northwest. I was hired to work in the Missoula Fire Department in September of 2001.

What is your favorite part of being a firefighter?

–*Joshua R. and Lauren H.*

Firefighter Kroll: Interacting with the people I work with and that it can be rewarding when you make a difference in someone's life.

Captain Knapstad: I make a living by helping people. How great is that?

Chief Painter: I love working as part of a team to help ease someone's suffering. Someone once said "They never call the fire department because something went *right*" and that is true. Still, even though the people who need our help are rarely glad to have to call us, we almost always can provide them with some sort of support in their time of need.

Joshua R.

Firefighter Campbell: I like the feeling of helping people in need. It is very satisfying knowing that you can make a personal difference in someone's life. The work schedule is also very flexible, which allows me to have fun outside of work.

Firefighter Cunniff: The best part about being a firefighter is knowing that you can help people in some of the worst times in their lives. We may have had three calls for car accidents in one shift already, but we have to remember this may be this person's first one. If there is a fire everything a person owns can get burned, and if we can stop the fire from destroying everything it really makes me feel good inside.

Firefighter Bennett: I love this job! I love the big, red trucks. I love the lights, the sirens, the calls, and the rush of not knowing what we will be doing when the call comes in. I love going to fires or rescues or car accidents. I love training and using big pieces of equipment to demolish cars and buildings. I love driving 60,000-pound ladder trucks. Racing through the streets at three o'clock in the morning when the rest of the city is asleep is just special. Seeing the sun come up after a long night of running calls is a unique experience. I love crawling into a burning building and not being able to see anything, yet I can feel the heat on my neck, in the creases of my coat, dragging a hose, and trying to locate the seat of the fire. I could go on and on. A lot of what we do is just flat fun.

Firefighter Brunell: The best part of my job is the knowledge that I am giving back to the community that I grew up in. I see people at their worst, due to fires, car wrecks, and accidents. My job is to help them with whatever has gone wrong, and I love knowing that I'm making their day better.

Jacob T.

How are you able to think of what to do in such a little amount of time?

—Jacob T.

Captain Knapstad: Training is a big part of how we react to different situations. If we prepare on the training grounds, it makes our job easier when "stuff" happens. Experience can also help, learning from what went right or wrong. Finally, when you make a decision and implement a plan, after a short time frame you need to reevaluate the plan to see if it's working. If it's not working, adjust or come up with an alternative plan.

Battalion Chief Scott: When I first became a battalion chief, I had a lot of check-off lists and would keep lots of notes. Now, after over four years in my position, I can usually remember the important things. I sit in a command vehicle outside the fire area. The captains on my shift are my eyes and ears inside the fire area and must keep me up to date on any situations that might need attention.

Chief Painter: While every single emergency is different (yet another reason this job always remains interesting), our goals are always the same: to keep people safe, to keep whatever problem exists from getting any worse, and to try to minimize damage. When firefighters are not responding to emergencies, they spend a great deal of time training (refreshing their knowledge, working with their equipment, practicing the different tactics they will use on a broad range of emergencies). We have written guidelines for most types of emergencies but after a few years, the firefighters usually have those guidelines deeply ingrained in their brains.

Firefighter Campbell: This very thing is one trait they look for in firefighters: You must be decisive! Training is a big part of that as well. We train on many things over and over. That way, when the real call comes in, we can act quickly and safely.

Firefighter Bennett: One of my favorite quotes is by Aristotle, and he said, "We are what we repeatedly do. Excellence then, is not an act, but a habit." This pretty much explains how we remember things. We train constantly. We do the same tasks over and over again. This breeds familiarity, confidence, muscle memory, and the ability to function under stress. By doing something 30, 40, 50, or 100 times, then when the pressure is on, a lot of the actions required are second nature. Thinking is still present but much faster.

Practice, practice, practice.

What is your greatest fear?

—Milly H. L.

Firefighter Kroll: Getting burned!

Captain Knapstad: My biggest fear is to have another fireman that I'm responsible for get injured or killed.

Battalion Chief Scott: Losing my air supply inside a smoke-filled structure. Now that my job doesn't usually require me to be inside a burning structure, I worry about the firefighters' safety under my command.

Chief Painter: I hope to never have to have one of my occupational brothers or sisters seriously hurt or killed at an emergency. Our focus, first and foremost, is on the safety of our personnel. But many of the things we are asked to do are inherently dangerous and unpredictable. We tell each other each day to "Stay low" and "Be careful out there."

Firefighter Bennett: The first is perhaps the most obvious. I don't ever want to not go home to my family at the end of my shift. I accept the fact that this may be asked of me someday, and if so, so be it. However, just thinking of not being there to see my kids grow up and doing the things I love with them is brutal.

My second greatest fear is the fear of failure. I fear failing my brothers and sisters. I fear failing the public who I have been entrusted to serve. And, lastly, I fear failing myself. I need to be the best firefighter I can be. My fellow firefighters depend on me for that.

Firefighter Brunell: My greatest fear is that I won't be able to reach someone in time to save his or her life.

What do you do when you get scared?

–Divya T.

Firefighter Kroll: I rely on the people I am working with to help me through it.

Battalion Chief Scott: I rely on my training and a very optimistic attitude.

Chief Painter: One of my favorite authors, Robert A. Heinlein, wrote "Courage is the complement of fear. A man who is fearless cannot be courageous. He is also a fool."

Divya T.

If I ever get to the point where I am not afraid of some of the things firefighters are asked to do, then I should find something else to do. Firefighters should have a healthy respect for and a little fear of fire. The key is to be able to act in the face of that fear, and the best way to learn to do that is to practice the things firefighters do when they are called to such emergencies.

Firefighter Cunniff: I try to take a deep breath, think, and look at how I can remedy the situation.

Firefighter Campbell: Sometimes I pray. You have to be strong in the difficult situations and keep your wits about you.

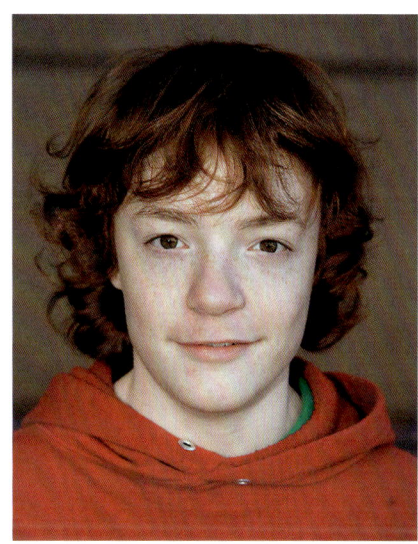

Josh C.

Firefighter Bennett: Fortunately, firefighters are very rarely alone in our tasks. If I am crawling through a burning house, I will have at least one more person with me. And if I am scared at that point, chances are that other individual is scared, too! It helps knowing that I have team members who are more than likely going through the same emotions I am.

What is the hardest part of saving people's lives every day?

–Josh R.

Firefighter Kroll: Fortunately, we don't have to save someone's life every day. There are a lot of people that we help that may just be hurt or sick and need us to help them in other ways. Making a difference in someone's outcome is very rewarding, so it isn't hard at all.

Battalion Chief Scott: Some times you have to put yourself in dangerous situations to save a life.

Chief Painter: Well, believe it or not, even despite the efforts of Missoula's firefighters, police, ambulance, and hospital workers, people still are injured and die. That's hard enough, but I would say the most difficult thing for most of our firefighters (me included) is when we see children suffer.

Firefighter Cunniff: The hardest part is not knowing what is going to happen when you get to a scene and that sometimes when you arrive there is nothing you can do.

Firefighter Bennett: There are many things that make this a job that can be hard to do. We see people at their worst, in the worst moment of their lives. Witnessing suffering on a daily basis can wear a person out over time. Not being able to save a life despite our best efforts takes some getting used to.

Firefighter Brunell: The hardest part is the knowledge that we can't save everyone. Unfortunately we see death in our job.

On a day when you are tired and very out of it what keeps you motivated?

−Haley G.

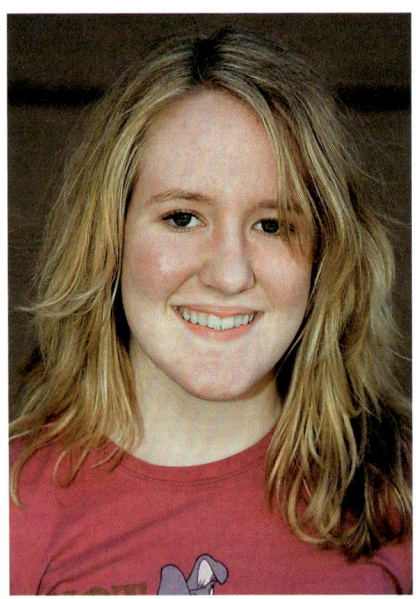

Haley G.

Firefighter Kroll: I know I have talked a lot about my co-workers, but we are kind of like a big team. If someone is a little down, that creates a target for someone else to take advantage of, and there are a lot of practical jokers around our station. There seems there is always someone that is willing to lighten the mood up for you if you are a little down that day.

Battalion Chief Scott: Coffee and my extended family here at the fire station.

Chief Painter: Two things that help pick me up physically when I am at work and need some motivation are exercise (all of our fire stations now have equipment for that purpose) and listening to music. It is especially helpful to do both at the same time. If that doesn't help pull me out of my doldrums, I find some firefighters willing to tell me some stories about combating entropy over a cup of coffee.

Firefighter Cunniff: Staying motivated does take its toll, especially when you don't have any fun calls for a while and you get lulled into a rut. What keeps me motivated is knowing that the biggest fire in my life could be five minutes away.

Firefighter Bennett: Perhaps I need to read a good article, talk to a fellow firefighter, or even run a few calls to get motivated. Very often I can be dragging and then I will look out the window as we go up the street and realize that I could have a job that I dislike or be stuck in an office somewhere. That will usually shake me out of it.

Firefighter Brunell: Knowing that the next alarm could be the alarm where something goes wrong and someone gets hurt or killed.

Firefighter Campbell: Knowing that you have the best job in the world!

CHAPTER 7

Virtual Apprentice
FIREFIGHTER FOR A DAY

What's a firefighter's favorite word? Fire? Nope. It's prevention. You're fire marshal for the day. Following is a to-do list that you can work on by yourself, with a couple friends, or, if your teacher agrees, as a classroom assignment.

8:00 Go online to the Web site of the United States Fire Administration's Kid Pages at http://www.usfa.fema.gov/kids. Browse the information you find there to learn all you can about fire safety. When you think you're ready click on http://www.usfa.dhs.gov/kids/parents-teachers/quizzes.shtm and print copies of the escape planning, home fire safety, and smoke alarm quizzes. How many of these questions can you answer correctly? If you miss any of the correct answers, go back and find out why.

9:00 When you think you've got enough fire safety know-how to become an official junior fire marshal go back to http://www.usfa.dhs.gov/kids and "Become a Jr. Fire Marshal." Test your fire safety knowledge by taking the Junior Fire Marshal certificate quiz. Celebrate the successful completion of this task by following the instructions to print out an official Junior Fire Marshal certificate.

10:00 Use all this new fire safety knowledge to conduct a thorough inspection of all the rooms of your house. Where are the smoke detectors? Are they working correctly? Make a list of any potential fire hazards that you find.

11:00 Make a fire safety chart showing all the ways your family can get out of your house in the event of a fire. Be as specific as possible and make sure that your plan includes escape routes for every member of the family in every room of the house.

12:00 LUNCH!

1:00 Conduct a mock fire drill with your family. Or make arrangements with your teacher to conduct one with your class at school. Take note of what goes right and what goes wrong. Of course, the goal is for everyone to respond in a cool, calm, and confident manner. Keep in mind that element of surprise is an important indicator of how prepared everyone is to get out quick.

2:00 Based on what you observed during the mock fire drill, write an official letter to your family or class that includes your suggestions for ensuring their safety in the event of a fire. Be sure to include comments about all the things they did right and recommendations for improvement. If necessary, include charts or maps to show them the best fire exit routes.

3:00 You've been invited to make a presentation about fire prevention to a group of students your own age. What do you think they need to know? How can make the information interesting and memorable? Can you think of any fun jingles or a rap tune to help key points stick in their minds? They say that when you have a choice it's better to show and tell in presentation. How can you get your audience involved in the presentation? Can you make a fire safety poster or put together a Powerpoint presentation that will help emphasize important fire safety rules.

4:00 Give a five-minute presentation about fire safety. At the end ask your teacher (or a trusted friend) to give you feedback. What were the best parts of the presentation? Are there any things you need to work on to improve? How could you change the presentation for a younger audience (like a group of kindergartners)?

5:00 Back to the fire station to shine your boots, clean your equipment, and get ready for another day of keeping your world and the people you care about safe from fires.

Virtual Apprentice

FIREFIGHTER FOR A DAY: FIELD REPORT

If this is your book, use the space below to jot down a few notes about your Virtual Apprentice experience (or use a blank sheet of paper if this book doesn't belong to you). What did you do? What was it like? How did you do with each activity? Don't be stingy with the details!

8:00 FIRE SAFETY WEB SURFING: _____
_____
_____
_____

9:00 JUNIOR FIRE MARSHAL: _____
_____
_____
_____

10:00 HOME INSPECTION: _____
_____
_____
_____

11:00 FIRE SAFETY CHART: _____
_____
_____
_____

12:00 LUNCH: _____

_____

_____

1:00 MOCK FIRE DRILL: _____

_____

_____

_____

2:00 FAMILY FIRE SAFETY LETTER: _____

_____

_____

_____

3:00 FIRE SAFETY POSTER: _____

_____

_____

_____

4:00 FIRE SAFETY PRESENTATION: _____

_____

_____

_____

5:00 FIREFIGHTER PREPARATION: _____

_____

_____

_____

Count Me In (or Out)

FUTURE FIREFIGHTER OR FREAKED OUT?

Hats off to the men and women on the big red trucks! Maybe, like them, firefighting is in your blood. Maybe not. Here are a few questions to help you sort it out. If this is not your book, don't even think about writing in it! Use a blank sheet of paper to record your answers.

The best part of my Virtual Apprentice day was: _____

_____

_____

Being a firefighter ❏ is ❏ isn't as easy as I thought it would be because:

_____

_____

_____

In case of emergency, look for me

❏ In the thick of things where I can do the most good.

❏ On the sidelines rooting for the good guys and gals.

❏ As far away as possible.

The best way to describe me in a pinch is

❏ Cool, calm, and collected. For example: _____

_____

❏ Totally freaked out. For example: _____

_____

_____

The idea of my becoming a firefighter strikes me as

❏ A good idea because: _____

_____

❏ Totally ridiculous because: _____

_____

Other "hot" jobs in firefighting that I find interesting are:

_____

_____

_____

Something I didn't know about firefighting until I read this book is:

_____

_____

_____

Something else I'd like to learn about firefighting is:

_____

_____

_____

As for a future in firefighting

❏ Sign me up. Here's what I need to do next: _____

_____

_____

❏ Forget about it. Here's what I'd like to do instead: _____

_____

_____

APPENDIX

More Firefighting Resources

BOOKS

Fan the flames of your firefighting interest by reading books such as:

Budd, E.S. *Rescue Machines at Work: Fire Engines.* Eden Prairie, Minn.: Child's World, 2000.

Gibbons, Gail. *Fire! Fire!* New York: HarperCollins, 1984.

Gorrell, Gena K. *Catching Fire: The Story of Firefighting.* Platsburgh, N.Y.: Tundra Books, 1999.

Gottschalk, Jack. *Firefighting.* New York: DK Publishing, 2002.

Hanson, Anne E. *Fire Engines.* Mankato, Minn.: Capstone Press, 2001.

Kelley, Alison Turnbull. *First to Arrive: Firefighters at Ground Zero.* Philadelphia, Penn.: Chelsea House, 2003.

Levine, Ellen. *If You Lived at the Time of the Great San Francisco Earthquake.* New York: Scholastic, 1987.

Tanaka, Shelley. *A Day That Changed America: Earthquake!* New York: Hyperion, 2004.

PROFESSIONAL ASSOCIATIONS

Professional organizations like these are where you'll find late-breaking firefighting news and all kinds of useful information:

International Association of Fire Chiefs
4025 Fair Ridge Drive
Suite 300
Fairfax, Virginia 22033-2868
http://www.iafc.org

International Association of Fire Fighters
1750 New York Avenue NW
Washington, D.C. 20006-5395
http://www.iaff.org

National Association of State Fire Marshals
1319 F Street NW
Suite 301
Washington, D.C. 20004
http://www.firemarshals.org

National Fire Protection Association
1 Batterymarch Park
Quincy, Massachusetts 02169-7471
http://www.nfpa.org

National Interagency Fire Center
3833 South Development Avenue
Boise, Idaho 83705-5354
http://www.nifc.gov

National Smokejumpers Association
P.O. Box 1022
Lakeside, Montana 59922-1022
http://www.smokejumpers.com

National Volunteer Fire Council
1050 17th Street NW
Suite 490
Washington, D.C. 20036
http://www.nvfc.org

INDEX